Old Victoria

A Copper Mining Ghost Town

Ontonagon County

Mikel B. Classen

Yooper History Hunter Series

Modern History Press

Ann Arbor, MI

Additional artwork by Joanna Walitalo.

Book #1 in the Yooper History Hunter Series
Learn more at www.MikelBClassen.com

ISBN 978-1-61599-819-7 paperback
ISBN 978-1-61599-820-3 hardcover
ISBN 978-1-61599-821-0 eBook

Modern History Press   www.ModernHistoryPress.com
5145 Pontiac Trail   info@ModernHistoryPress.com
Ann Arbor, MI 48105   toll free 888-761-6268

Distributed by Ingram Book Group (USA, CAN, UK, EU, AU)

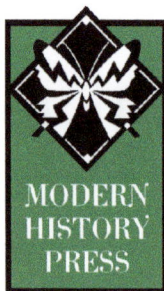

MODERN
HISTORY
PRESS

Contents

Photos in this Book ii

Introducing: The Yooper History Hunter Series 1

A Taste of Homestead Upper Peninsula.................................... 2

The Legends of Old Victoria Begin.. 4

The Victoria Mine and a Village Grows Up.............................. 6

Accidents and a Hard Life.. 15

Death and Resurrection 19

Victoria Area Points of Interest 23

About the Author... 24

Photos in this Book

A Pair of Cabins at Old Victoria...2

Historical Victoria: New Homes (1914) 3

Old Victoria Cabin – Visitor Center...4

Replica of the Ontonagon Boulder 5

Old Victoria - Victoria Mine Site 6

Old Victoria - Hoist Building Remains 7

Victoria 1899 Mine 8

Constructing the Taylor Hydraulic Air Compressor.....................9

In the Compressor - 375 feet underground. 9

Blow Off from the Compressor 10

Blow Off from the Compressor – Frozen in the Winter.............11

Compressor Explosion Aftermath (1914) 13

Locomotive Powered by Compressed Air Storage Tank 13

Victoria Schoolhouse Renovation (1903).................................14

Victoria Renovated School 15

Victoria Miners with Candles to Light their Way.....................16

Victoria Miner Riding the Skip to the Bottom 17

Miners with Candles and Controversial Compressed Air Drill ...18

Old Victoria – Restored Log Cabin 20

Old Victoria - Mining Captain's Residence 21

Old Victoria Dam...22

O-Kun-De-Kun Waterfalls 23

Ontonagon Lighthouse...24

Introducing: The Yooper History Hunter Series

The Yooper History Hunter Series is a new concept in history books by Mikel B. Classen, award winning historian and Modern History Press. These full-color books have a short format so they can be read quickly and priced lower. Part travel information combined with a comprehensive history of the subject matter, the Yooper History Hunter books focus on small individual subjects creating an indispensable collection of U.P. history.

Mikel shows the way that the reader can become a Yooper History Hunter by using contemporary and historical photography as well as directions to uncover these amazing remnants of Michigan's past. A classic hybrid of then and now, the reader becomes the explorer seeking out those ghostly remains of the past.

Mikel B Classen, Yooper history hunter himself, puts a modern day adventure between the pages of each book. He hands out the keys to unlocking the past while journeying through the present. After 40 years of researching history and exploring the Upper Peninsula with camera and pen, he reveals places and stories that made the U.P. what it is today, a place of fascination, beauty and significance.

Victoria Mine

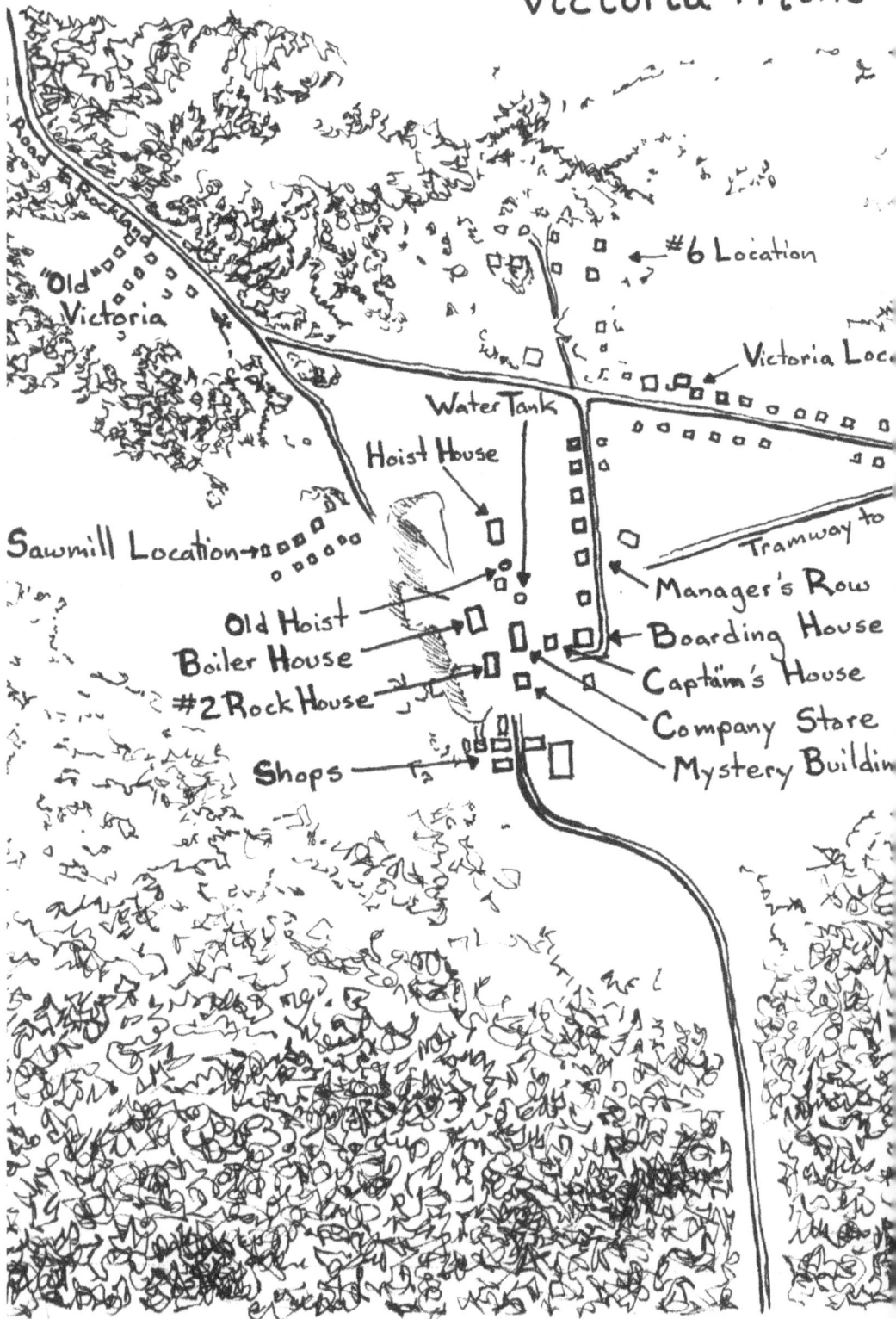

"Old Victoria"

Road to Rockland

#6 Location

Victoria Loc

Sawmill Location →

Water Tank

Hoist House

Tramway to

Manager's Row

Old Hoist

Boiler House

#2 Rock House

Boarding House

Captain's House

Company Store

Mystery Buildin

Shops

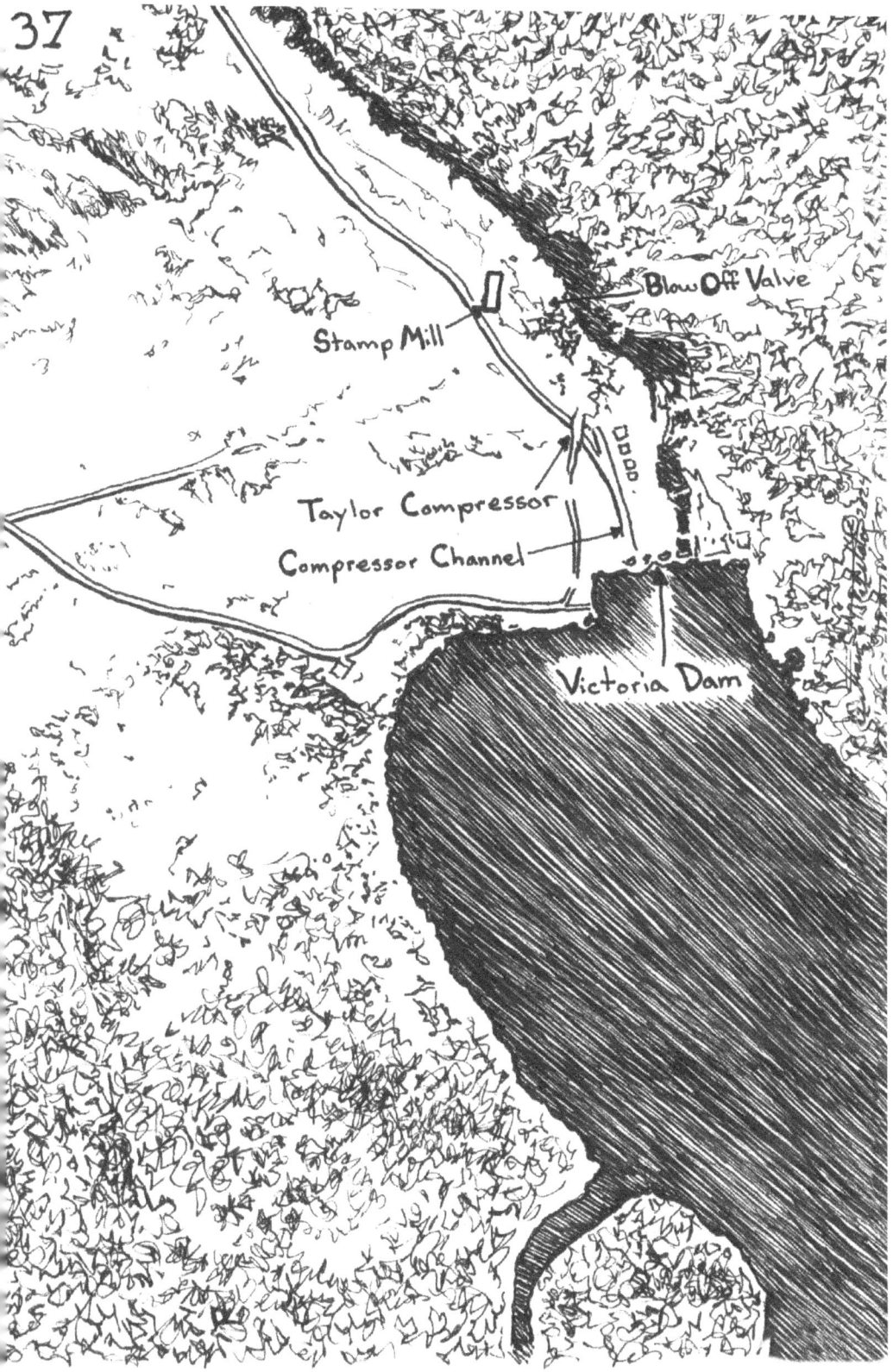

Blow Off Valve

Stamp Mill

Taylor Compressor

Compressor Channel

Victoria Dam

A Taste of Homestead Upper Peninsula

A Pair of Cabins at Old Victoria

Old Victoria, a ghost town from the copper boom, shows what life was like homesteading in Michigan's Upper Peninsula. Over the years, some of the site has been destroyed or collapsed; still, many of Old Victoria's original homesteads still stand. Thanks to the efforts of a local group, The Society for the Restoration of Old Victoria, quite a few of the buildings have been restored and refurnished in their original condition. Unlike Fayette, the U.P.'s best known ghost town and a small shipping port on Lake Michigan, Victoria is a remote, rugged mining town, buried in the Ontonagon wilderness.

Thus, Victoria is one of the least known – yet most interesting – of the Upper Peninsula's attractions. The town was carved out of one of the harshest sections of the rugged U.P. landscape. Situated at the top of a Michigan mountain, part of the picturesque Ontonagon River Gorge, Victoria is within the Gogebic Mineral Range. When visiting, you get the feel for what it was like to struggle in a remote mining town.

Historical Victoria: New Homes (1914)

Old Victoria is two miles west of Rockland on Victoria Road. The approach drive is steep and rocky, showing the formidable and rough landscape—the forest has regrown to much of its original density. Modern lifestyles make it difficult to relate to these early pioneers' survival. People of today might find it difficult to imagine the hardships involved in making a living or, for some lucky few, a fortune from the minerals buried beneath.

The restored homesteads try to recreate that life. Walking through them, you see how small and constricted they were. The old and crude furniture and the ancient wood-burning stoves prove that comfort was rare. Life was spartan, at best.

Old Victoria, part of the Keweenaw Heritage Trail, is listed on the National Register of Historic Places. Like so many locations in the U.P., Old Victoria's history was fascinating—even legendary—long before the arrival of the Europeans. Along the shore of the Ontonagon River lay the Ontonagon Boulder, a two-ton piece of copper that the Native Americans worshiped and the Europeans lusted for.

Old Victoria Cabin – Visitor Center

The Legends of Old Victoria Begin

As early as the late 1600s, explorers dug into the hill that made up Victoria's landscape, looking for copper. Alexander Henry, the British adventurer and explorer, sojourned there to search for the Ontonagon Boulder. Finding it, he chipped off a large piece; then, he had his companions begin digging into the high banks of the Ontonagon hoping to find a rich vein. Next, Henry left for Sault Ste. Marie to get supplies to begin mining in earnest. When he returned, his men were waiting for him: the diggings had collapsed from soft clay. They called it quits right then and there (the full story of the Ontonagon Boulder can be found in my

Alexander Henry (1739-1824)

book, *True Tales: The Forgotten History of Michigan's Upper Peninsula*).

A few years later, another early attempt was made to mine at the Victoria location. Building on the discovery of prehistoric copper culture pit mines in the ravines, the men again began digging at the Victoria location; however, because of the harsh winter, they abandoned the digging after only a few months. But they didn't leave without finding what they were looking for—the hill was rich with copper. At the bottom of their shaft was a one-ton piece of copper.

It lay there until the mid-1800s, when the copper boom caused speculators to open mines all over the U.P., and Victoria was no exception. Throughout the early years, the early mining company, then called the Forest Mining Company, invested hundreds of thousands of dollars on various mining ventures, with no payoffs. The mine, perched at the top of a mountain, presented many problems difficult to overcome. The solutions to these problems gave Victoria its notable place in history.

Replica of the Ontonagon Boulder
Courtesy: Ontonagon County Historical Society Museum

Old Victoria - Victoria Mine Site

The Victoria Mine and a Village Grows Up

The area was not going to make getting at the copper easy. From the start, mining attempts were plagued with problems. Serious mining operations began, but logistics kept the mining business from seeing a profit. The Forest Mining Company sold its holdings, and the Victoria Mining Company came into being. Named after Queen Victoria, it illustrated the influence of the English immigrants who flocked to the upper Michigan mines.

Then, a forest fire destroyed all the buildings Victoria Mining Company had erected. Its stamp mill (a mill for crushing ore for further extraction) was destroyed, and, to prevent fire danger, a new one was built in a clearing along the Ontonagon River, only to have it swept away by a spring flood.

Old Victoria - Hoist Building Remains

So much time was spent overcoming natural obstacles that little mining was done. The company even erected a giant windmill to pump water out of the mine and into the stamp mill. The Victoria Mining Company erected a sawmill to create timbers for shaft supports and used wood to fire the steam boilers, but, because of the forest fire, wood was scarce. Coal seemed like it could be a great alternative fuel, but the grade was so steep from nearby Rockland that a railroad was ultimately ruled out.

The Ontonagon River wasn't bridged, so a ferry had to be utilized, raising the price of coal from $2.50 a ton to $8.00, just because of the four-mile journey. At the turn of the century, the mine still made no money. Something had to be done.

One answer came in the form of mining Captain Thomas Hooper. Born in Cornwall, England, he cut his teeth on the surrounding mines of the Ontonagon region. Working his way up the mine worker ladder, he eventually became a very efficient mining captain, the name given to mining supervisors. He was hired to bring the Victoria to profitability. It was a daunting task, at which all others had previously failed.

Hooper was innovative and he looked closely at what he had to work with. He decided the Ontonagon River could be a help and not the hindrance it had been in the past. There was a large waterfall, which Hooper decided to dam. Now, he needed something to harness the river's power.

Thomas Hooper (1842 - 1920)

Victoria 1899 Mine

Constructing the Taylor Hydraulic Air Compressor

That something came in the form of a device called the Taylor Hydraulic Air Compressor. In 1899, Hooper began a new effort to profitably extract the copper from Victoria, paying close attention to the lessons of the past and making decisions that would allow the company to be run as inexpensively as possible. The compressor would let them do that.

In the Compressor - 375 feet underground.

Blow Off from the Compressor

Blow Off from the Compressor – Frozen in the Winter

Building the compressor was a massive project, requiring a cavern to be dug 400 feet underground—a mining project unto itself. Three large shafts had to be dug to take in the water from the Ontonagon River. The tailings were pulled out by rock buckets (called skips), with men riding them in and out. At one point, a skip cable snapped, fell and hit another one, killing four men and injuring several others.

The compressor worked by dropping the river's water nearly four hundred feet into the ground through three intake shafts. Large amounts of air were introduced into the water through a special apparatus containing numerous small tubes located over each intake. The countless bubbles were then released in the air chamber, cut from solid rock. The air was trapped by water at both the intake and the outlet and by the solid rock of the chamber itself. The air main then bled off the compressed air from near the top of the air chamber.

Because the compressor worked solely off the river's water, the compressed air was icy cold, creating a strange sight: icicles would hang from the exhaust valves, and the operators had to bundle in heavy coats on hot summer days, working around the frost-covered machines. The icicles that formed were dozens of feet high, surrounding a blow-off pipe that sprayed water and air into the sky. It was a bizarre sight that became known as the "Victoria Geyser;" locals recall their ancestors telling them about it. To this day, the compressor still exists, and some believe it is still in working order.

At one point, in 1916, the blowoff valve, a pressure release, froze. The pressure buildup was so great that it backed up into one of the intake valves and exploded. The force of the explosion drove a 3,100-pound piece of iron up and out of the shaft, destroying a building that was covering the intake.

Compressor Explosion Aftermath (1914)

Because of the compressor, suddenly the mine had no fuel costs. One of the mining mechanics created a compressed air locomotive to move ore to the mill on the river. As far as anyone knows, it is the only one ever made. For the first time, the mine became profitable. A dam was built to direct the water into the compressor. The town grew and eventually thrived, becoming a company town— with the company owning the houses.

Locomotive Powered by Compressed Air Storage Tank

Being a "company town" meant the company owned everything: the houses and the company store, which basically led to the company owning the workers. Anything needed was provided by the company and any money owed was taken directly from the paychecks. It was a great system for the company; not so much for the workers. They were underpaid, making it hard for them to get out of debt.

Victoria Schoolhouse Renovation (1903)

Victoria was a good-sized community with over 60 homes, a church, store, doctor's office and school. Finns, Swedes and Cornish immigrants made up most of the town's population. As the town grew, there came a need for expansion, and the school was lifted up to add an extra floor so that the lower floor became the top floor and the new one became the first floor. It seems a little backward, but it apparently worked.

Victoria Renovated School

Accidents and a Hard Life

Mining isn't an easy job and the threat of getting killed deep in some dark hole constantly loomed. Mining accidents resulting in death were not unusual and at the Victoria Mine, things were no different. Twenty-five men lost their lives during the time of the mine's operation, the highest number in Ontonagon County. A miner at the Victoria Mine had a 1 in 6 chance of not surviving the job. At one point, seven fatal accidents occurred in one year.

Beginning in April 1907, accidents were frequent—too frequent. On April 4, two men were swinging picks on a rock pile when one of the men struck a drill hole that still contained some blasting powder. Only one died—the one that hit the powder. On April 25, a miner was walking when his candle blew out, plunging his surroundings into pitch black. He fell down a shaft in the darkness and died.

Victoria Miners with Candles to Light their Way

On May 2, another man walked into a shaft, unable to see because of dust from blasting. Another accident occurred on June 1. A miner, who had just finished his lunch, stood up and walked into a shaft. He fell 100 feet and was found on the tenth level hanging from a timber. He was still alive, but his skull was crushed; he died shortly after.

On July 19, a man had his skull crushed when he was trying to place a timber support underground. So, 1907 was a rough year and one of the questions was, why were the men using only candles? With the water compressor, they could have easily powered electric lights in the mine. They never did that and, consequently, they had the highest death toll—by far than any

Victoria Miner Riding the Skip to the Bottom

other mine in the area. Although 1907 was a bad year, the accidents continued throughout the years Victoria Mines operated. 1908, 1910, 1912, 1913, 1914, 1915, 1916 and 1919 all recorded fatal accidents. There were two deaths before 1907.

Some of the other fatalities occurred around the rock skip, which was a big bucket that ran down into the mine. It was also the only way for the miners to descend to their work: riding it down to the lower levels. Occasionally, the system would break and fall, causing someone to get killed or pieces of rock headed for the surface would fall off and kill someone. Of all the mines in the same area, Victoria had the worst accident record. The Mass, the Adventure, the Minesota all had much better records, with the

Miners with Candles and Controversial Compressed Air Drill

Mass mine being closest, but still with half of the fatalities of Victoria. Along with this were dozens of injuries. Mining was a hazardous occupation, but, in Victoria, the danger was downright deadly. Since the company owned the houses, they could—if a miner was killed—displace his widow and children, making room for family that had an able-bodied worker. Of course, a widow would be paid a stipend for her loss and was allowed a month to stay before being evicted.

Copper prices fell at the end of World War I, and the town began its slide into a memory. The Victoria Mine slowly shut down in the early 1920s. The last family left Victoria in 1923, officially making it a ghost town.

It was the last copper mine in the region still operating. The surrounding mines had shut down when the prices fell. Victoria operated so cheaply that it was able to outlast the others in the area, but, finally, the falling prices forced them to give up. In 1928, the property was sold to the Copper District Power Company, and the buildings in the town eventually fell to decay or were torn down, and plans were announced to burn the ruins. Some area residents would hear nothing of that.

The Alexander Family in Victoria at their Home

Death and Resurrection

These residents created a group called the Society for the Restoration of Old Victoria and asked that the historic homes be spared. The society made its goal the complete restoration of the lower log locations and the marking out of the old mining building locations at the mine site. They created hiking trails, a picnic area and opened the old homes undergoing restoration, or have been restored to the public, at no charge. All who come to Victoria can see a little of the lives of those pioneer miners who lived there so long ago.

Old Victoria – Restored Log Cabin

Old Victoria is believed to be one of the oldest log-cabin villages, in its original location, in the United States. In recent years, a history class from one of the area schools, conducted a heritage project there. School children actually took residence in the village, living the lives of the pioneers, thereby learning about the past through experience. Restoration is carried out by means of donations and fund-raising activities conducted by the society. Some of these are geared toward the general public and can enhance a visit to Old Victoria. One example, Log Cabin Day, is in mid-June; you can catch re-enactments and traditional cooking.

An annual craft fair held at Old Victoria is another such event. It features artists and craftspeople from all over. Held in August, the fair features traditionally cooked foods. The woodstove-cooked cinnamon rolls are gaining a reputation. Artists' tables are inexpensive.

Old Victoria - Mining Captain's Residence

In the future, the society hopes to start restoring more buildings. Campsite creation is being considered as well; currently, there are none. The coming years promise to be ones of change in Old Victoria. Overwhelming scenery and special history make a visit worthwhile. The Society for the Restoration of Old Victoria is looking for any help it can get and encourages everyone to join.

If you are interested in any information on Old Victoria, its activities, or events write:

Old Victoria Historical Site,
25401 Victoria Dam Rd.,
Rockland, Michigan 49960

Old Victoria Dam

Located about a mile from the Old Victoria Restoration, this dam and hydroelectric plant on the west branch of the Ontonagon River date back to the early 1930s, when they provided power for mining and logging industries in the western Upper Peninsula. Victoria Dam is also home to a waterfall produced by the dam's overflow channel that varies depending on season and rainfall.

For more information about Old Victoria ghost town:

- Old Victoria (U.S. National Park Service)
  www.nps.gov/places/old-victoria.htm
- Old Victoria Keweenaw Heritage Site
  www.superiortrails.com/old-victoria.html
- FaceBook Old Victoria
  www.facebook.com/oldvictoria

If you enjoyed this story, you might enjoy my book, *True Tales: the Forgotten History of Michigan's Upper Peninsula.*

Victoria Area Points of Interest

After exploring the Victoria ghost town, find an adventure in these other places of local interest.

O-Kun-De-Kun Waterfalls: Named after local Native American chief, located approximately eight miles north of Bruce Crossing on U.S. 45.

Rockland Museum: 40 National Ave (US-45), Rockland, MI 49960 (906) 886-2821

Adventure Mine Tours: 200 Adventure Ave., Greenland, MI 49929

Ontonagon Historical Museum: 422 River Street, Ontonagon, MI 49953, (906) 884-6165, www.ontonagonmuseum.org

Ontonagon Lighthouse Tours: (run through the Ontonagon Museum)

Porcupine Mountains Wilderness State Park: 15 miles west of Ontonagon. Take M-64 to Silver City, go straight on 107 to Porkies.

Ontonagon Lighthouse

Tours of the lighthouse are available 7 days a week from 11:00 a.m. to 3:00 p.m. from mid-May through mid-October and in the off-season by appointment, weather conditions permitting (906-884-6165). Those wishing to take the tour can buy tickets at the Ontonagon County Historical Museum (located at 422 River Street) or at the lighthouse (take River Road off M-64 and follow the signs). Admission is $5.00 per person (18 and over).

About the Author

Mikel B. Classen has been writing and photographing northern Michigan in newspapers and magazines for forty years, creating feature articles about the life and culture of Michigan's north country. A journalist, historian, photographer and author with a fascination of the world around him, he enjoys researching and writing about lost stories from the past. He is founder of the U.P.

Reader and is a member of the Board of Directors for the Upper Peninsula Publishers and Authors Association. In 2020, Mikel won the Historical Society of Michigan's, George Follo Award for Upper Peninsula History.

Classen makes his home in the oldest city in Michigan, historic Sault Ste. Marie. He is also a collector of out-of-print history books, and historical photographs and prints of Upper Michigan. At Northern Michigan University, he studied English, history, journalism and photography.

His books, *Au Sable Point Lighthouse, Beacon on Lake Superior's Shipwreck Coast,* was published in 2014 and *Teddy Roosevelt and the Marquette Libel Trial,* was published in 2015 by the History Press. His book of fiction, *Lake Superior Tales,* won the 2020 U.P. Notable Book Award. *Points North: Discover Hidden Campgrounds, Natural Wonders, and Waterways of the Upper Peninsula,* published in 2019, which has received the Historical Society of Michigan's, "Outstanding Michigan History Publication," along with a 2021 U.P. Notable Book Award. Since then, he has released *True Tale:, the Forgotten History of Michigan's Upper Peninsula,* and *Faces, Places, & Days Gone By: a Pictorial History of Michigan's Upper Peninsula.* Mikel is co-author of the *Yooper Ale Trails* along with Jon C. Stott. All of these books were published by Modern History Press. In late 2023 Mikel released his first novel, *The Alexandria Code: an Isabella Carter Adventure,* published by Modern History Press.

To learn more about Mikel B. Classen and to see more of his work, please visit his website **www.mikelbclassen.com**

What Were Pioneer Days *Really* Like in the U.P.?

The combination of mining, maritime and lumbering history created a culture in the U.P. that is unique to the Midwest. Discover true stories of the rough and dangerous times of the Upper Peninsula frontier that are as enjoyable as they are educational. You'll find no conventional romantic or whitewashed history here. Instead, you will be astonished by the true hardships and facets of trying to settle a frontier sandwiched among the three Great Lakes.

These pages are populated by Native Americans and the European immigrants, looking for their personal promised land-whether to raise families, avoid the law, start a new life or just get rich... no matter what it took. Mineral hunters, outlaws, men of honor creating civilization out of wilderness and the women of strength that accompanied them, the Upper Peninsula called to all. Among the eye-opening stories, you'll find *True Tales* includes:

- Dan Seavey, the infamous pirate based out of Escanaba
- Angelique Mott, who was marooned with her husband on Isle Royale for 9 months with just a handful of provisions and no weapons or tools
- Vigilantes who broke up the notorious sex trafficking rings-- protected by stockades, gunmen, and feral dogs--in Seney, Sac Bay, Ewen, Trout Creek, Ontonagon and Bruce Crossing
- Klaus L. Hamringa, the lightkeeper hero who received a commendation of valor for saving the crews of the Monarch and Kiowa shipwrecks
- The strange story of stagecoach robber Reimund (Black Bart) Holzhey

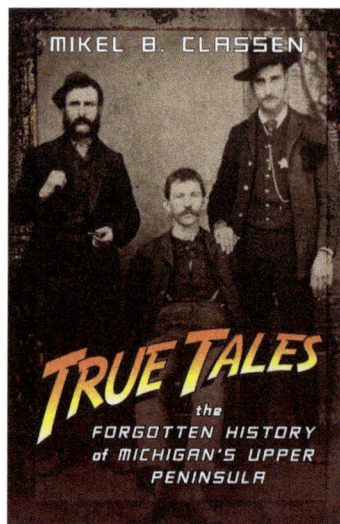

From Modern History Press

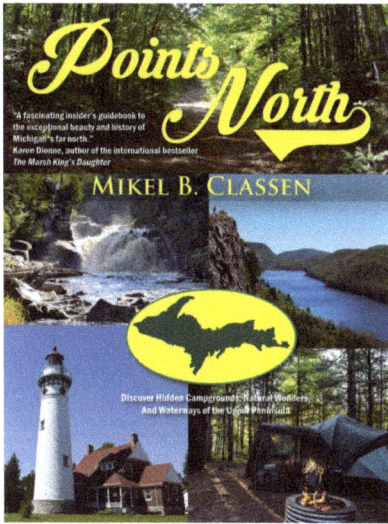

Featuring more than 150 color photos of the U.P.'s most beautiful, historic, and natural locations

I've spent many years exploring the wilderness of Michigan's Upper Peninsula (U.P.), and one thing has become apparent: no matter what part in which you find yourself, fascinating sights are around every corner. There are parks, wilderness areas and museums. There are ghost towns and places named after legends. There are trails to be walked and waterways to be paddled. In the U.P., life is meant to be lived to the fullest. In this book, I've listed 40 destinations from every corner of the U.P. that have places of interest. Some reflect rich history, while others highlight the natural wonders that abound. So, join in the adventures. The Upper Peninsula is an open book--the one that's in your hand.

"Without a doubt, Mikel Classen's *Points North* needs to be in every library, gift shop and quality bookstore throughout state. Not only does Classen bring alive the U.P. through his polished words, his masterful use of color photography also makes this book absolutely beautiful. *Points North* will long stand as a tremendous tribute to one of the most remarkable parts of our country."
--Michael Carrier, author, *Murder on Sugar Island*

"Mikel Classen's love for Michigan's Upper Peninsula shines from every page in *Points North*, a fascinating insider's guidebook to the exceptional beauty and history of Michigan's far north. Whether you're still in the planning stages of your trip, or you're looking back fondly on the memories you created--even if you wish merely to enjoy a virtual tour of the Upper Peninsula's natural wonders from the comfort of your armchair, you need this book."
--Karen Dionne, author, *The Marsh King's Daughter*

Learn more at **www.PointsNorthBooks.com**

From Modern History Press